Domestication and Feed Conversion Ratio of Cardisoma Armatum, the Nigerian Land Crab

Jolaosho Toheeb

Bibliographic information published by the German National Library:

The German National Library lists this publication in the National Bibliography; detailed bibliographic data are available on the Internet at http://dnb.dnb.de.

ISBN: 9783346343918
This book is also available as an ebook.

Print and binding: Books on Demand GmbH, Norderstedt, Germany
Printed on acid-free paper from responsible sources.

GRIN web shop: https://www.grin.com/document/985062

Table of content

CHAPTER ONE

INTRODUCTION

1.1 BACKGROUND OF THE STUDY

Crabs are part of the basic components of the ecosystem and they are consumed as food in many countries. Over 100 species of crabs are known worldwide with nine species common in West African countries especially Nigeria.

Crabs are decapods crustaceans which have a very short tail and are covered with a thick shell, or exoskeleton and are armed with a single pair of claws. There are over 6,793 species of crab spread across the oceans, fresh water, and even on land. Among the species are the land crab (*Cardisoma Armatum*), the big fisted swim crab (*callinectes amnicola*), and (*callinectes latmanus*). These 3 species are edible ones (Abby-kalio 1982, Hart and Chindah, 1998).Crabs mostly occur at the mouth of estuaries and along the course of many main rivers (Oyekanmi 1984). Crabs which are the basic components of the ecosystem are the most advanced members of the phylum Arthropoda.

The freshwater crabs of Nigeria are true crabs which can be distinguished from false crabs by not having 5 pairs of the pereiopods totally or partly concealed beneath the carapace, the antennae were always placed between the inner margin of orbit and fused Pterygostomial region with endotome. True crabs belong to the Suborder Brachyura of order Decapoda under Class Crustacea. It shows the greatest size range of all arthropods such as observed in lobsters, prawns, crayfishes, shrimps, hermit crab and true crabs. (Abby-kalio 1982)

Crabs are widely distributed in the tropical and temperate regions of the world. Literature information is scarce on the food values of most edible crabs in West Africa despite the richness of these gastropods. Most people rate the fish higher in preference to crabs which are considered inferior and food for the low income earners. Over 100 species of crabs are known Worldwide with two species; *Sudanonautes africanus* and *Cardisoma armatum* readily known to exist with economic value in West Africa; the mud crabs (*Sudanonautes africanus*) found in estuaries and mangrove. *S. africanus* inhabits cracks and holes when fully matured and as small individuals are found under rocks in the littoral zone. (Herklots 1988 and Rathbun.1986)

Crab constitutes one of the main sources of animal protein most especially among coastal dwellers in some parts of Nigeria. Crabs have a high ash, mineral and crude fiber content, serves as a source of minerals consumed either wholly and partially by sick folks, and is often recommended for pregnant women. Crab fisheries, can be capture or farmed and are found in all of the world's oceans. There are many aquatic crabs, semi-terrestrial crabs and terrestrial crabs, particularly in tropical regions, due to their nutritional value and abundance scoop of inquisitiveness in nutritional values of crabs are generated in recent times, they are highly exploited but only their fleshes are consumed. Animal protein is very vital in the diet because of the various functions it performs. Fish, beef, pork and poultry products are some of the good sources of animal protein which are used for the growth and repair of body tissues. In developing countries, the high cost of these highly valued animal proteins (i.e. fish, beef, and poultry) has made it impossible for the less privilege to eat them. However, lesser valued animal such as crabs may be used to cater for the protein needs of the less privileged people. Generally, Land crab (*Cardisoma Armatum*) is sometimes referred to as the rainbow crab in Africa, known as moon crab or patriot crab in Nigeria. According to (Burggren and Mchahon 1988) Land crabs is defined as the crabs that show significant behavior, Morphological, physiological and/or biochemical adaptations per-mitting extended activities out of the water.

However gecarcinid species is a terrestrial crab, as all of them must return to the set for larvae release (cuesta et al. 2002) .The known zoeal of all gecarcinids species are marine planktonic, and complete larvae development consists of five or six stages followed by a megalopa.

Crabs constitute an important species in the traditional prawn or fish culture system in some coastal state and union territories, and have become increasingly popular by virtue of meat, quality and large size.

Cardisoma Armatum belongs to the taxonomy of Dormain: Eucaryota, Kingdom: Animalia, Subkingdom: Metazoa and Eumetazoa, Grade: Bilateral, Branch: Protostomia, Infrakingdom: ecdysozoa, Phylum: Arthropoda, Subphylum: Crustacea, Class: Malacostraca, Superorder: Eucarida, Order: Decapoda, Suborder: Pleocyemata, Infraorder: Reptantia, Section: brachyuran, Superfamily: Grapsoidea, Family: Gecarcinidea, Subfamily: Cardisoma, Species: *Cardisoma armatum*

Land crab (*Cardisoma Armatum*) is an omnivore which means it eats both plant and meat. In nature, the crab constitute a nuisance by damaging set of net in water and it diet consist mainly of vegetable and other small insects, reptiles, Amphibians and fish, though crabs also serve as prey to mammals, birds and fishes but they constitute one of the most important members of estuarine food chain. In the current feeding method frozen red worms are used for feeding them. *Cardisoma* uses it claws to collect and hold the food and direct it toward its head part for eating.

Recreational fishermen seek enormous interest in this animal for recreational purposes and Many physiologist have used crabs as experimental animals because it is readily available, economic viable, hardiness and complex life cycle (Smallegange and Vander Meer 2003) Adult are sometimes labeled as soap dish crabs in pet industry. This name derives from aggressive nature as adults and when being ship to pet stores, they are often in soap dishes to prevent them from killing each other. The names moon crab and soapdish crab are sometimes applied to other similar crab species, leading to frequent confusion with other colorful crabs which are the three (3) remaining species of *Cardisoma*.

These species originates from coastal regions of western Africa, but it also occurs inland along some deltas (e.g. the Volta river delta), and on islands such Cape Verde. When young, these crabs typically have a bluish/violet carapace, red-colored legs, and whitish claws. This colouration usually fades as the crab grows older. They can reach a carapace size of 20cm across, although captive's individuals rarely reach this size. Their diet often consists of mainly fruit, vegetation and carrion. They are known to be cannibalistic, and will consume smaller crabs, smaller reptiles and amphibians, mollusks, fish and insects if they can catch them while the juvenile and adult crabs spend most of their time on dry land, the females must return to the ocean to release their eggs. The eggs hatch into microscopic larvae, and later on developed they will drown. This is a land crab and cannot remain submerged for long periods. .(Herklots and Rathbun.1986)

Crabs patterns of distribution are related to salinity, sediment characteristics, temperature, ions and water availability, debit and pH and topography. Associated biotic factors like litters' floristic diversity, forest complexity and harvesting were identified as the determining factors which affected the distribution and abundance of mangrove crabs.(Herklots and Rathbun.1986)

There are two major forms of land base *cardisoma armatum* aquaculture which are fattening of crabs with low flesh /water and grow out of juvenile to market size.

Grow out system can be pond base with or without mangroves, although inter tidal pond can also be used .pond of 0.5-1 acre is suitable for crab culture .The maximum stocking density maybe on crab per square meter .The pond culture period extend from three to six months. The water quality needed to be maintained at salinity of 10-35ppt,temperature of 28-32°c,dissolved oxygen of 5ppm-7ppm,ph 7.8- 8.5,water depth of 0.5-1.0m,however this type of culture is not very popular in some countries like India due to non availability of juvenile/seed crab in sufficient quantity and lack of good quality artificial feed.

Crab fattening is essentially an operation during which post molts or water crabs are kept in captivity for short period of 20-30Days until they flesh out or immature female crab are held until their gonads develop and fill the mantle cavity. This is very popular throughout the Asia countries due to increase demand for gravid female and large size hard shelled crab in sea food restaurant. (Herklots and Rathbun.1986)

Crab seeds are available in the nature of all sizes. Juvenile crabs can be collected from estuaries, lakes, mangroves and salt water lagoons by using bamboo traps, lift nets and scissor net. Land crab culture depends on mainly on natural feed supply and therefore is a major limiting factor for culture of crabs. (Herklots and Rathbun.1986)

1.2 OBJECTIVE

The objective of this study is to evaluate the domestication and the growth performance of the land crab (*Cardisoma Armatum*) cultured in a rectangular concrete unit.

CHAPTER TWO

LITERATURE REVIEW

2.1 DESCRIPTION

Cardisoma Armatum originates in the western coastal of Africa but also occur inland along some deltas and on islands such As Cape Verde. (Anon 2007)

However, most crabs are just as happy to hide. In the wild they would dig burrows. In captivity they're just as likely to scoot under a rock or piece of driftwood. They live at the edge of the water. They survive under water for long periods but need access to land to catch their breath. Like all crabs, they have gills that must be kept moist, so they need access to water at all times. A tank with half water and half land works well. Or you can give them an easy to enter and exit dish of water and mostly land. *Cardisoma Armatum* was originally shipped in plastic soap dish containers and was thus called "soap dish crabs." They came in yellow, purple, red, and blue. They have a long list of known names in different, like rainbow crabs, moon crabs or patriot crabs, Columbian crabs, and others. Land crabs are scavengers; they eat whatever they find including live plants. (Anon 2007)

When young, these crabs typically have bluish/violet carapace, orange-red coloured legs, light –colored abdomen and whitish claws. The colouration usually fades as the animal grows older and they can reach a carapace size of 20cm across, although captive individuals rarely reach the size. The female's abdominal apron is wider facial field with mandibles. During mating these crabs so like to do cartwheels while Male crabs have narrow abdomen and their claws are of different sizes (Anon 2007)

2.2 LIFE HISTORY AND POPULATION BIOLOGY

The location and time of copulation is not known but has been observed outside their burrow (Taissoun, 1974). The sperm are carried by the female and fertilization of the eggs is internal (Gifford 1962). Fertilized eggs are carried externally on the abdomen .Egg mass is related to the body weight with 300g females carrying approximately 300,000-700,000 eggs (Gifford, 1962, Feliciano 1962, Taissoun 1974). Females lay their fertilized eggs in salt water along the coast. This occurs with lunar phases (nights preceding and after the full moon). The spawning season varies somewhat with geographic location ; In south Florida it is July- October (Gifford, 1962) on Andros the season frequently ends by September and in Venezuela the season continues until later in the year (Taissoun. 1974). During the early evening there are mass migrations of ovigorous females to lays their eggs in coastal saltwater. Female crab may spawn several times during the season. The eggs immediately hatch into free swimming larvae.

Larvae development proceeds through five zoeal and one megalopa stage (costlow and bookhout, 1968a). On land, crabs dig burrows to the watertable (salt or fresh) and must moisten their gills regularly (Gifford, 1962). Larger crabs tend to move further inland and dig burrows up to a maximum of 2.0m (Feliciano, 1962; Herried and Gifford, 1963).

However, where the terrain is low such as Venezuela (Taissoun, 1974) and Andros, marketable size crabs (over 350g) are commonly found in shallow (less than 2 ft) burrows. There are normally only one crab per burrow and one opening per burrow (Feliciano, 1962; herried and Gifford, 1963; Taissoun 1974).

Crabs deposit their solid wastes at the entrance of their burrows which makes it easy to determine if the burrow is "active". Juvenile and adult crabs can live with either salt or freshwater (depending on the water table) in the burrows but larvae will develop only in salt water (Gifford, 1962).

2.3 CONDITIONS REQUIRED FOR CRAB CULTURE INVOLVED THE FOLLOWING

2.3.1 THE SOIL QUALITY

The suitable soil suitable for crab fattening is sandy or sandy clay as a sandy bottom discourage burrowing.

2.3.2 THE WATER QUALITY

Abundant of good quality water should be available. Crabs are highly tolerant salinity conditions, so brackish water would be ideal for crab culture operation. The water salinity should be about 15-30ppt and not less than 10ppt,the ph should be around 7.8-8.5,temperature at 28°c-30°c and dissolve oxygen of more than 5ppm.

2.3.3 THE SIZE OF POND AND CONSTRUCTION

Crab culture is practiced in ponds, cages, or pens. Small tidal ponds ranging from 0.025 to 1hac in size with water depth of 0.5-1.0m are generally use for the purpose; a sandy bottom is preferable to discourage burrowing. Bunds should have minimum width of 1.0m at the top to prevent crabs from escaping through the bunds. Crabs are capable of climbing over the bunds, which can be prevented by fixing and hanging fences on the dykes.

2.3.4 WATER SUPPLY AND DRAINAGE

In tidal crab, Water exchange is through tides where sluices gates are used to regulate the inflow and outflow of the tidal water. The sluices gates are fitted to prevent the escape of crabs in region where tidal influence is less. However, in artificial culturing ponds, brackish or seawater or freshwater is pumped in and drained out through outlets.

2.3.5 POND MANAGEMENT

Pond preparation for crab culture is done by draining out the water and allowing it to dry before liming is done. The process will help to get rid of the obnoxious gases and also the pathogen and microbes which are detrimental to the cultured crabs. The water is later pumped in from the reservoir tank to maintain height of above 1.5m. (Baliao, D.D., De Los Santos, M.A. & Franco, N.M. 1999)

2.3.6 STOCKING

Soft shelled crab of 8cm carapace width and above or crabs of more than 550g are stocked and usually the stocking density of 1-3 crabs/m² is usually practiced. Crabs of similar sizes are preferred to reduce cannibalism. Pond can be divided into compartment so that crabs of the same sizes can stored together to avoid cannibalism, Male and female crab can also be stored separately and can be cultured along with fishes like chinos, and mullets which are practiced by farmers in India , however it is not recommended. Monoculture of crabs based on scientific practices yield better result.

The size at which crablets are stocked into mangrove pens depends on the size of the netting used and also if a nursery is installed within the pen. In the case of a pen with a nursery, small crablets of less than 1cm CW can be stocked. Once the crablets are larger than 2cm they can be released into the main pen. The density at stocking that is used should produce 1-1.5tonnes/ha.This figure needs to takes into account typical mortality rates experienced, target production per hectare and the size of crab required at harvest. (Baliao, D.D., De Los Santos, M.A. & Franco, N.M. 1999)

2.3.7 TRANSPORTATION OF CRABLETS

Crablets older than larvae can be transported with or without water. Cooling crablets before transport is recommended to prevent both moulting and lower oxygen consumption during transport. Crablets can also be transported in water using the same method as used for post-larval. (Baliao, D.D., De Los Santos, M.A. & Franco, N.M. 1999)

2.3.8 STOCKING OPERATIONS

Preferably, all stock should arrive at a farm with health checks already completed. Even with a clean bill of health, the quality of crablets (and juvenile stages of other species being stocked for polyculture) must be assessed for quality prior to stocking.

To assess the quality of a batch of crablets one should examine the following criteria:

• Visual health – picks a subsample of crablets and examine for fouling, unusual coloration, damage to legs or claws.

•Size variation – while crablets in any batch may be at different moult stages, size variation should be minimal. Extreme variation in size indicates batches may have been combined. Too large a variation in size increases the likelihood of losses to cannibalism.

•Activity – if crablets have been transported to farm at temperatures less than optimal, they may be sluggish. After equilibrating to ambient temperature, they should be actively walking or swimming shrimp or juvenile fish. Out of water, crablets can be transported in containers on moist sand or damp cloth, in containers that are lined and covered to minimize evaporation and resulting desiccation, while ventilated to ensure they can respire adequately. It is recommended that transport of crablets out of water should not exceed 30 hours. (Baliao, D.D., De Los Santos, M.A. & Franco, N.M. 1999)

2.3.9 FEEDING AND FEEDING METHODS

Cardisoma armatum is an omnivore crabs which means it preys on both plant and meat. In nature, the crab diet consists mainly of vegetables and other lives such as small insects, reptiles, amphibians, fishes and it also feed on small quantities of detritus and plant materials such as banana leaves, onions and some other forage leaves. In current feeding method frozen red worms are used for feeding them. *Cardisoma armatum* uses its claws to collect and hold the food and direct it toward its head part for eating the food. (Matt Clarke 2005 -11-14).

Feeding rates utilized are often quoted as a percentage of the body weight of crabs in the pond as the quality and nutritive value and other natural feeds varies significantly; these figures are difficult to generalize on. Feeding frequency is typically once or twice a day to minimize the risk of cannibalism, which is considered more likely if the crabs are hungry. (Matt Clarke 2005 -11-14).

However banana leaves {Musa spp. (Musaceae)} is a food crop that generates large amount of forage materials that can be used to feed livestock. Banana leaves which grow continuously from the center of the stem are broad blades,1-4m long x 0.7-1.0m wide, with a pronounced supporting mid rib. Banana leaves and pseudo stem can be fed to animal in flesh, ensiled, or dried form. They are bye product of banana production and are usually available near to the field and processing plants. They can found in all tropical and sub region of Asia, America, Africa and Australia where banana are grown. Banana foliage is a useful source of roughages in many tropical countries .Particularly, it can be used as an emergency feed in case of draught or feed shortage. (Scot C. Nelson et al: 2006).

Banana production yield large quantity of forage biomass, for an average crop fruiting 1.5 times in a year. Forage biomass can amount to 13t/ha/year. Banana leaves contain about 85% of water and 10-17% protein (DM basis). Pseudo stems contains mostly water(92-95%)and very little protein(30-35% DM) fiber content is high, in the 50-70% range for NDF and about 30-45%DM for ADF. They can be fed fresh or sun dried whole or chipped. Pseudo stem are easily ensiled if chipped and mixed with easily with fermentable sources of carbohydrates such as molasses or rice barns. (Scot C. Nelson et al: 2006).

2.4 MUSA SPECIE (BANANA)

Banana is an edible fruit, botanically berry, produced by several kinds of large herbaceous, flowering plants in the genus Musa. In some countries, Bananas used for cooking may be called plantains. The fruit is variable in size, color, and firmness, but is usually elongated and curved with soft flesh richen starch covered with a rind which maybe green, yellow, red, purple, or brown when ripe. The fruits grow in clusters hanging from the top of the plant. Almost all modern edible *parthenocarpic* (seedless) bananas come from two wild species – *Musa acuminata* and *Musa balbisiana.* The scientific names of most cultivated bananas are *Musa acuminata* , *Musa balbisiana*, and *Musa paradisiacal* for the hybrid *Musa acuminate* and *M. balbisiana* , depending on their genomic constitution. The old scientific name *Musa sapientum* is no longer used. Musa species are native to tropical Indomalaya and Australia, and are likely to have been first domesticated in Papua New Guinea. They are grown in at least 107 countries, primarily for their fruit, and to a lesser extent to make fiber, banana wine, and banana beer and as ornamental plants .Worldwide; there is no sharp distinction between "bananas" and" plantains". Especially in the Americas and Europe,"banana" usually refers to soft, sweet, dessert bananas, particularly those of the Cavendish group, which are the main exports from banana-growing countries. (Scot C. Nelson et al: 2006.)

By contrast, Musa cultivars with firmer, starchier fruit are called "plantains". In other regions, such as Southeast Asia, many more kinds of banana are grown and eaten, so the simple twofold distinction is not useful and is not made in local languages. (Scot C. Nelson et al: 2006).

The term "banana" is also used as the common name for the plants which produce the fruit. This can extend to other members of the genus *Musa* like the scarlet banana (*Musa coccinea*), pink banana (*Musa velutina*) and the Fe'I bananas. It can also refer to members of the genus *Ensete*, like the snow banana (*Ensete glaucum*) and the economically important false banana (*Ensete ventricosum*). Both genera are classified under the banana family, *Musaceae*. (Scot C. Nelson et al: 2006.)

2.5 WATER TEMPARATURE

As a tropical species, *cardisoma* adult are generally not heavy impacted by low winter temperature, in the wild, they probably face much more variation in their ambient temperature. However *cardisoma* larvae are highly affected by water temperature below 20°c. (Gifford1962, Feliciano 1962, Taissoun 1974).

2.6 PHYSICAL TOLERANCE

Terrestrialization in crabs is accompanied by a decrease in the number, volumes and area occupied by gills. This adaptation perhaps decreases the amount of evaporative surface area, thus minimizing the effect of desiccation. The gill structure in land crabs also shows increased sclerotization to support platelets which may result in decreased permeability of gases. (Gifford, 1962, Feliciano 1962, Taissoun 1974).

Oxygen availability is lowered in terrestrial crabs. A possible compensation for this potential problem is to increase the folds of the vascular epithelium and vascular tufts associated with the gills. The gills of *cardisoma* are relatively small, less than 10% of the volume of the branchial cavity. However, gills surface area is increased due to extensive folding and vascularization of the epithelial sheet lining the branchial cavity. *Cardisoma*, when tested against other crabs, both terrestrial and aquatic, maintained a nearly constant rate of oxygen consumption in both air and water. This was accomplished by its manipulating the rates of oxygen extraction and ventilation over the gills. In water, where oxygen concentration arte lower, and extraction rates tends to fall off, *Cardisoma* increases its ventilation rate. In air where both oxygen concentration and extraction rate are higher, *Cardisoma* decrease its ventilation rate. *Cardisoma Armatum* is limited by mechanisms of water conversion. They are unable to conserve water; thus though they are able to function extremely well in moist air, they die when faced with desiccating conditions for over 2days. (Gifford, 1962, Feliciano 1962, Taissoun 1974).

2.7 BEHAVIOUR CHARACTERS

Cardisoma Armatum are aggressive towards conspecies as well as basically any living being smaller in size(or rather, that fit in the claws), Male crab tend to be more territorial than females and will defend burrows and small surrounding territories. When the crabs argue, they try to tear off each other's arms and legs. If they lose any parts, they are at a disadvantage when eating and when fighting. Separate any walking (or non-walking) wounded. They will grow new parts at their next molt. It may take half a year for them to get back into fighting trim. They often sit motionlessly on some wood above the waterline or in their hole on the land part for hours or even for days. They like to burrow (in nature they dig over two meter deep tunnels down to the ground water).Also like to sit in "knee-deep" water so that their eyes are above the water line and they can breathe easily. (Taissoun 1974).

Aquatic and land plants are hardly ever eaten or shredded, However the crabs might uproot them when walking over or undermine land plants when digging. They like to have their back carapace stroked with a toothbrush when they are very tame and also their claws. They recognize their keeper and can be fed with tweezers or even by hand. By judging the vibrations of steps they can discern between a person they know and a stranger, (Taissoun 1974).

2.8 REPRODUCTION

The reproduction cycle is closely linked to seasonal weather patterns and lunar phase. Heavy rain in the spring initiates migrations of crabs. At this time, *Cardisoma* gains weight rapidly as foraging intensity is increased for the first few weeks of the migratory period. Male crabs actively court ripe females during this period. Fertilization of eggs is internal and throughout July and August, most females carry external egg masses. Eggs are carried for approximately 2 weeks prior to hatching and must be released into saltwater for larvae to survive. Females typically complete spawning migrations within 1-2 days and generally spawn within 1-2 days of full moon. Thus, though *cardisoma armatum* and other terrestrial crabs have been successful invaders of the land, they are still heavy dependent on the ocean for at least part of the life cycle. Several spawns per year may occur, when reproducing, the crabs return to the beach, and the females releases their larvae into the sea after A good two weeks, there larvae would have go through different and several zoea and Megalopa stages before leaving the sea and go back to live on land (at Alfred Wegener institute for polar and marine research in Bremerhaven). (Smallegange, I.M. and Van Der Meer, J. 2003)

2.9 HABITAT DISTRIBUTION

Cardisoma Armatum originates from coastal region of western Africa, but it also occurs inland along some deltas and on islands such as carp Verde. (Anon 2007).

The natural habitat of the crab can be moist sandy areas, mangroves, under houses, in areas adjacent to sources of brackish or sea water or inland areas of larger islands. Adult crabs excavate and construct deep burrows for housing purposes. These burrows are usually a part of a large colony where burrows intersect. Generally only a single individual occupies a burrow. However small juvenile less than 10mm do not dig their own burrows, and will often share a burrow with adult crabs, (Taissoun 1974).

2.10 LOCOMOTION

Cardisoma are typically able to walk at a rate of 100m/hr and may move over 500m or more per evening, However if actively foraging, this distance is decreased to an average of less than 100m.

Land crabs orient themselves by using polarized light during the day, and by identifying the brightest sector of the horizon at night. Other cues such as geotaxis, substrate vibration, land marks and prevailing winds may also aid in orientation, especially during migratory periods (Gifford, 1962).

2.11 SALINITY

Adult *cardisoma* utilize a range of habitat and tolerates salinities from Zero to hypersaline because they are usually surround by air rather than water, they are essentially closed systems and not subject to salt gain on or through the body surface.

Furthermore, they are able to osmoregulate down to fresh water levels. Optimal salinities range for larvae development is 20-40ppt.Larvae survival fell off dramatically at salinities below 15ppt and above 45ppt. At 10ppt, Zoea developed only to the second stage and at 45ppt less than 1% of larvae completed development of the megalopa. (Gifford, 1962)

2.12 CATCHING METHOD

Cardisoma are actively sought and captured for private consumption and commercially sale throughout their range in Caribbean, Central and South America. In most location crabs are primarily captured by hand (With the aid of stick, glove or net) when they are intercepted outside their burrows during spawning season.

When crabs are harvested during the remainder of the year, they are captured from their burrows by hand reaching into burrows (Taissoun 1974) or with the aid of traps (Feliciano 1962). Hand catching in burrow is restricted to relatively shallow burrows. Small traps that are set over individual burrows are limited to catching one crab per set; however, these traps can be set more once per day. Mass impoundment type traps are dependent on baiting (e.g. mangoes in South America) to attracts crabs.

2.13 STOCK CONTROL NETTING

Unlike marine shrimp, Land crabs can leave water and spend considerable periods of time on land. As a result, if a barrier of some type did not surround a crab aquaculture pond, stock crabs would be able to walk out of the pond, which would be a direct financial loss to the farmer. To counter this mud crab behavior, netting typically surrounds crab culture ponds and netting height may vary from 20 to 50 cm in height above the top of the pond. The netting is typically supported by posts and may be topped with plastic. The plastic topping is added as mud crabs are good climbers and they can climb up netting, but are unable to climb up clear plastic sheeting. A variation in design for the stock control netting is that it can be constructed along the inner top slope of the pond, at such an angle that a crab would have to hang upside down to climb over the net to escape from the pond. For anything larger than a juvenile size crab, this is usually not possible. However, from a maintenance perspective, netting around the top of the pond is easier to maintain, as access is simpler for staff. Working inside the inner top slope of the pond can cause unnecessary erosion to the bank of the pond depending on its construction and soil type. (Dat, H.D. 1999.)

2.14 HARVEST TECHNIQUES

While crabs are reasonably tolerant to a wide range of environmental variables, it is recommended that unnecessary stress be avoided during harvest. Wherever possible, avoid harvest activities during high temperatures, typically experienced during the day. In addition, preliminary grading at harvest should be undertaken as quickly as possible and unsuitable mud crabs returned to the pond to complete grow-out. Once some of the crabs in a pond are identified as large enough to harvest, harvesting of crabs can commence. Crab pots of various designs can be used to harvest crabs. These are baited with food attractive to crabs. It is found that the largest crabs in a pond tend to enter traps first. As this is the case, ponds can be partially or selectively harvested on a regular basis, progressively removing the larger crabs from the pond. To complete the harvest, either trapping is continued until no more crabs are trapped, or the pond is drain harvested, with crabs collected from the pond's drain or the lowest part of the pond. (Dat, H.D. 1999.)

2.15 POST-HARVEST

Once harvested from an aquaculture facility, the crabs must be examined, cleaned and stored for transportation to a processing facility, unless this is situated on-farm. Recent work in Australia has highlighted improvements that can be made to enhance the post harvest survival of crabs through a better understanding of their physiology and stresses that they are exposed to throughout the supply chain. On harvest, farmers should tie the claws of crabs with string or vine to ensure that they cannot damage one another or people involved in further processing, distributing and selling the product. (Dat, H.D. 1999.)

If crabs are to be transported to a temporary storage facility or processing shed, once taken from the water they are typically stored in a fish crate or container and covered with dampened material to reduce desiccation and protect against flies. Some preliminary sorting (by size, sex or missing limbs) may be undertaken at this stage depending on arrangements with the facility that will be doing the final processing and packaging. If Land crabs are to be held for some time before going to a processing facility, the containers holding the crabs should be kept in the shade and kept moist to avoid dehydration and heat stress. A spray system can be used to keep crabs moist, temperatures moderate and humidity high. Post-harvest mortalities can be high (4–

10 percent) if crabs are harvested in localities remote from markets and supply chains stress crabs significantly. (Dat, H.D. 1999.)

2.16 MUSA SPP. (BANANA)

2.16.1 DESCRIPTION

The banana plant is the largest herbaceous flowering plant. All the above-ground parts of a banana plant grow from a structure usually called a “corm". Plants are normally tall and fairly sturdy, and are often mistaken for trees, but what appears to be a trunk is actually a "false stem" or pseudo stem. (Scot C. Nelson et al: 2006.) Bananas grow in a wide variety of soils, as long as the soil is at least 60 cm (24 in) deep, has good drainage and is not compacted. The leaves of banana plants are composed of a "stalk" (petiole) and a blade (lamina). The base of the petiole widens to form a sheath; the tightly packed sheaths make up the pseudo stem, which is all that supports the plant. The edges of the sheath meet when it is first produced, making it tubular. As new growth occurs in the centre of the pseudo stem the edges are forced apart.

Cultivated banana plants vary in height depending on the variety and growing conditions. Most are around 5 m (16 ft) tall, with a range from 'Dwarf Cavendish' plants at around 3 m (10 ft) to 'Gros Michel' at 7 m (23 ft) or more. Leaves are spirally arranged and may grow 2.7 meters long and 60 centimeters wide (8.9 ft × 2.0 ft wide).They are easily torn by the wind, resulting in the familiar ragged frond look.(Scot C. Nelson et al: 2006.)

When a banana plant is mature, the corm stops producing new leaves and begins to form a flower spike or inflorescence. A stem develops which grows up inside the pseudo stem; carrying the immature inflorescence until eventually it emerges at the top. Each pseudo stem normally produces a single inflorescence, also known as the "banana heart". (More are sometimes produced; an exceptional plant in the Philippines produced five. After fruiting, the pseudo stem dies, but offshoots will normally have developed from the base, so that the plant as a whole is perennial. In the plantation system of cultivation, only one of the offshoots will be allowed to develop in order to maintain spacing. The inflorescence contains many bracts (sometimes incorrectly referred to as petals) between rows of flowers. The female flowers (which can develop into fruit) appear in rows further up the stem (closer to the leaves) from the rows of male flowers. The ovary is inferior, meaning that the tiny petals and other flower parts appear at the tip of the ovary. The banana fruits develop from the banana heart, in a large hanging cluster, made up of tiers (called "hands"), with up to 20 fruit to a tier. The hanging cluster is known as a bunch, comprising 3–20 tiers, or commercially as a "banana stem", and can weigh 30–50 kilograms (66–110 lb). (Scot C. Nelson et al: 2006.)

Individual banana fruits (commonly known as a banana or "finger") average 125grams (0.276 lb), of which approximately 75% is water and 25% dry matter (nutrient table, lower right).The fruit has been described as a "leathery berry". There is a protective outer layer (a peel or skin) with numerous long, thin strings (the phloem bundles), which run lengthwise between the skin and the edible inner portion. The inner part of the common yellow dessert variety can be split lengthwise into three sections that correspond to the inner portions of the three carpel's by manually deforming the unopened fruit. In cultivated varieties, the seeds are diminished nearly to non-existence; their remnants are tiny black specks in the interior of the fruit. (Scot C. Nelson et al: 2006.)

Bananas are naturally slightly radioactive, more so than most other fruits, because of their potassium content and the small amounts of the isotope potassium-40 found in naturally occurring potassium. The banana equivalent dose of radiation is sometimes used in nuclear communication to compare radiation levels and exposures. (Scot C. Nelson et al: 2006.)

2.16.2 TAXONOMY

The genus Musa was created by Carl Linnaeus in 1753. The name may be derived from Antonius Musa, physician to the Emperor Augustus, or Linnaeus may have adapted the Arabic word for banana, *mauz*. Musa is in the family *Musaceae*. The APG III system assigns *Musaceae* to the order *Zingiberales* , part of the *commelinid clade* of the monocotyledonous flowering plants. Some 70 species of Musa were recognized by the World Checklist of Selected Plant Families as of January 2013; several produce edible fruit, while others are cultivated as ornamentals.). (Scot C. Nelson et al: 2006.)

The classification of cultivated bananas has long been a problematic issue for taxonomists. Linnaeus originally placed bananas into two species based only on their uses as food: *Musa sapientum* for dessert bananas and *Musa paradisiaca* for plantains. Subsequently further species names were added. However, this approach proved inadequate to address the sheer number of cultivars existing in the primary center of diversity of the genus, Southeast Asia. Many of these cultivars were given names which proved to be synonyms. (Scot C. Nelson et al: 2006.)

In a series of papers published in 1947 onwards, Ernest Chessman showed that Linnaeus's *Musa sapientum* and *Musa paradisiaca* were actually cultivars and descendants of two wild seed-producing species, *Musa acuminata* and *Musa balbisiana,* both first described by Luigi Aloysius Colla. He recommended the abolition of Linnaeus's species in favor of reclassifying bananas according to three morphologically distinct groups of cultivars – those primarily exhibiting the botanical characteristics of *Musa balbisiana*, those primarily exhibiting the botanical characteristics of *Musa acuminata*, and those with characteristics that are the combination of the two. Researchers Norman Simmonds and Ken Shepherd proposed a genome-based nomenclature system in 1955. This system eliminated almost all the difficulties and inconsistencies of the earlier classification of bananas based on assigning scientific names to cultivated varieties. Despite this, the original names are still recognized by some authorities today, leading to confusion. Scot C. Nelson et al: 2006.

Musa spp, Plant genus of extraordinary significance to human societies, produces the fourth most important food in the world today (after rice, wheat and maize), bananas and plantains. *Musa* species grow in a wide range of environments and have varied human uses, ranging from the edible bananas and plantains of the tropics to cold-hardy fiber and ornamental plants. They have been a staple of the human diet since the dawn of recorded history. These large, perennial herbs, 2–9 m (6.6–30ft) in height, evolved in Southeast Asia, New Guinea, and the Indian subcontinent, developing in modern times secondary loci of genetic diversity in Africa, Latin America, and the Pacific. Musa species attained a position of central importance within Pacific societies: the plant is a source of food, beverages, fermentable sugars, medicines, flavorings, cooked foods, silage, fragrance, rope, cordage, garlands, shelter, clothing, smoking material, and numerous ceremonial and religious uses. With the exception of atoll islands, banana and plantain are ideally suited for traditional Pacific island agro forestry, for inter planting in diversified systems, and for plantation-style cultivation in full sun. Although mostly consumed locally in the Pacific region, the fruit enjoys a significant worldwide export market. (Scot C. Nelson et al: 2006).

2.16.3 Family: *Musaceae* (banana family)

Musa species (banana and plantain)

2.16.4 Distribution: Native to the Indo-Malesian, Asian, and Australian tropics, banana and plantain are now found throughout the tropics and subtropics. (Scot C. Nelson et al: 2006.)

2.16.5 Habitat: Widely adapted, growing at elevations of 0–920 m (0–3000 ft) or more, depending on latitude; mean annual temperatures of 26–30°C (79–86°F); annual rainfall of 2000 mm (80 in) or higher for commercial production. (Scot C. Nelson et al: 2006.)

2.16.6 Propagation: Banana and plantain are propagated principally by vegetative division and far more rarely by seeds (usually only for banana breeding, ornamental types, and wild species). In addition, tissue culture has become standard for commercial plantations in recent years, primarily because of the advantage of starting with disease-free planting material. Edible bananas are almost always seedless (however, some, such as 'Pisang Awak', produce many seeds when growing near a fertile pollen source. (Scot C. Nelson et al: 2006).

2.17 COMPOSITION OF MUSA SPECIE FOLIAGE (BANANA LEAF): The table shows the commercial composition and nutritional value of a banana leaf.

Main analysis unit Average SD Min Max Mb
Dry matter %dm 20.7 3.4 16.7 23.7 5
Ash %dm 11.4 1.4 9.5 13.7 14
Calcium g/kgdm 16.7 6.4 5.3 30.0 14
Ether extent %dm 5.6 2.1 1.1 8.7 14
Crude fiber %dm 28.6 3.8 19.1 34.8 14
Gross energy Mj/kgdm 18.1
Phosphorus g/kgdm 1.2 0.3 0.6 1.7 14
Potassium g/kgdm 25.1 8.9 14.9 50.5 13
Magnesium g/kgdm 3.6 0.8 1.9 4.5 13
Crude protein %dm 9.5 2.4 2.6 13.2 14

CHAPTER THREE

MATERIALS AND METHODS

3.1 Experimental site

150 crablets of *Cadisoma armatum* was used in this study at the Department of Fisheries Technology, Lagos State Polytechnic, Ishagamu road, Ikorodu, Lagos, Nigeria.

3.2 Sample Size and Technique

A total of 160 crablets (*Cardisoma Armatum)* was purchased from artisanal crab's farmer. The crablets (*Cardisoma Armatum)* were captured from the wild beside Majidun River Ikorodu Local Government, Lagos, Nigeria. The crablets were transported from Majidun where it was purchased to Lagos State Polytechnics where it's to be domesticated. A total of fifteen (15) units with a stocking rate of 10 crablets of varied size per treatment, making it a total of 150 crablets of *Cardisoma Armatum.*

3.3 Experimental animals and Design

Eight weeks feeding trial was conducted with 150 crablets of *Cardisoma armatum* of mixed sex breed and with average initial weight of between 10.0g and 10.5g to study the domestication and diet performance of *Cardisoma armatum* fed with Musa species foliage. The crablets was randomly allotted to five treatments with fifteen crablets serving as replicates. The experimental design that has been used is Complete Randomized Design (CRD). The crablets would be weighed fortnightly to determine their growth performance.

T1R1	T2R1	T3R1	T4R1	T5R1
T1R2	T2R2	T3R2	T4R2	T5R2
T1R3	T2R3	T3R3	T4R3	T5R3

An empty rectangular concrete pond of 388m length, 217.5m width and 94m depth were demarcated into fifteen (15) smaller rectangular units with 2x2 wood, wooden rod, using wire gauze and polythene (nylon) in between each unit. The polythene (nylon) to prevent the crablets from crawling out of the wire gauze to prevent them from moving from one unit to another due to the slippery effect of the polythene (nylon). Wire gaze, fine mesh net, wooden rod and 2x2 woods are used to construct the cover on the rectangular concrete units. Each unit has a length of 77.6m, 72.5m width while the depth remains the same for each unit (94m). The cover has four rectangular openings for easy accessibility.

Length per unit = 388m/5units =77.6m for a unit length,

Width per unit =217.5m/3units=72.5m for a unit width,

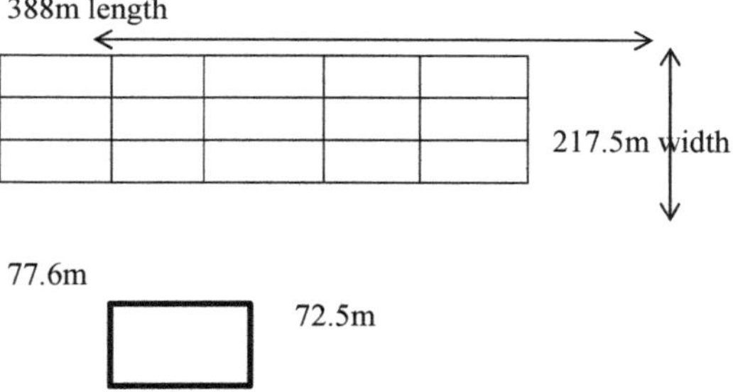

Each unit is filled with a mixture of sandy-clay soil at a ratio 50%:50% (1m depth).

3.4 Experimental diet

The fresh banana leaves Musa spp. foliage that was used for this experiment was freshly cut from banana tree from Lagos State Polytechnic farm. The foliage with stalk was weighed and feed to crablets at 5% body biomass.

3.5 Growth Parameter Matrix

Diet performance has been evaluated as follow: Mean weight gain (MWG), Specific Growth Rate (SGR), Feed Conversion Ratio (FCR) and percentage survival of the crablets.

3.6 Data Analysis

Data has been analyzed using species package. Analysis of variance produce for significant differences among treatment means. Means has been analyzed by Duncan's Multiple Range Test among individual weight at 0.05 significance level.

CHAPTER FOUR

RESULTS AND DISCUSSION

4.1 Data Presentation and Analysis

This study revealed the possibility of domesticating *Cardisoma armatum* fed with *Musa specie* foliage in rectangular concrete units.

Table 1: shows average weight gained by the *Cardisoma armatum* fed with banana foliage for 8 weeks in a controlled environment.

Table 1: Average Weight Gain and standard deviation

Week	T1	T2	T3	T4	T5	Mean	StdDev
1	158	167	163	171	175	**139.2**	**6.6483**
2	158	168	164	172	176	**140.0**	**6.9857**
3	158	173	171	178	178	**143.5**	**8.2037**
4	158	175	172	180	179	**144.7**	**8.8713**
5	158	164	162	173	167	**138.2**	**5.6303**
6	158	167	163	175	168	**139.5**	**6.3008**
7	158	111	112	163	119	**111.7**	**25.7158**
8	158	112	113	165	120	**112.7**	**25.7740**
Mean	**158.0**	**154.6**	**152.5**	**172.1**	**160.3**	**159.5**	**8.6886**
StdDev	**0.0000**	**26.8485**	**24.9743**	**5.8661**	**25.5217**	**12.7036**	

In table 1, The significant growth performance in Treatments 4 week 4 might be due to better consumption of food during these weeks which may be as a result of high amino acid in diet as banana leaves are rich in crude protein and calcium as reported by Scot C. Nelson and Randy C. 2006 who gave the proximate analyses of banana leave and posited that banana leaves contains 8.50% of Crude Protein and 1.04% of Calcium. Protein is the major nutrient that is require in largest amount for growth and development. Therefore, in nutritional study of crab protein requirement is given high priority because crab requires food that is very rich in protein according to (aqualandpetsplus.com; 2007).

The difference in growth pattern in all the treatments might be as results of climatic factors with the attendant of heavy rain and erratic temperature changes which can reduce the specific dynamic action of crabs this is in agreement with aqualandpetsplus.com; 2007

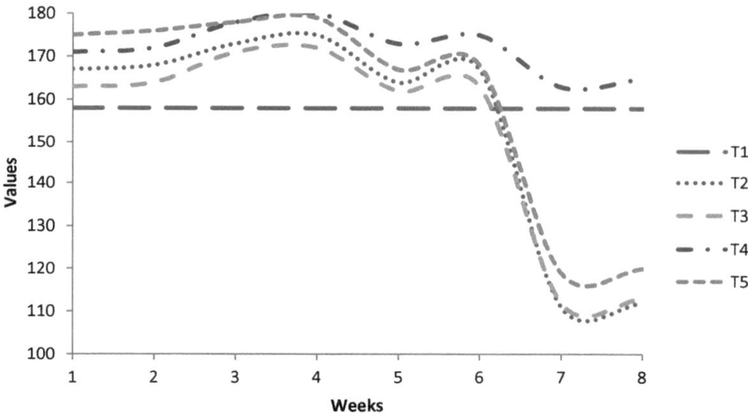

Figure 1: Line graph showing values of different treatments at different week

Figure 1 depicts that treatment one (T1) has the same values from week 1 to week 8. Treatment two (T2), treatment three (T3) and treatment five (T5) in week 1 have higher values than treatment one (T1) but went lower than treatment one (T1) at week 7 to week 8. Treatment four (T4) has the second highest value in week 1 but rise above all other treatments from week 3 to week 8 and still higher than T1 till the end of the 8th week. Treatment five (T5) increases at an increasing rate from week 1 to 4 week then decreases at week 5 to week 8.

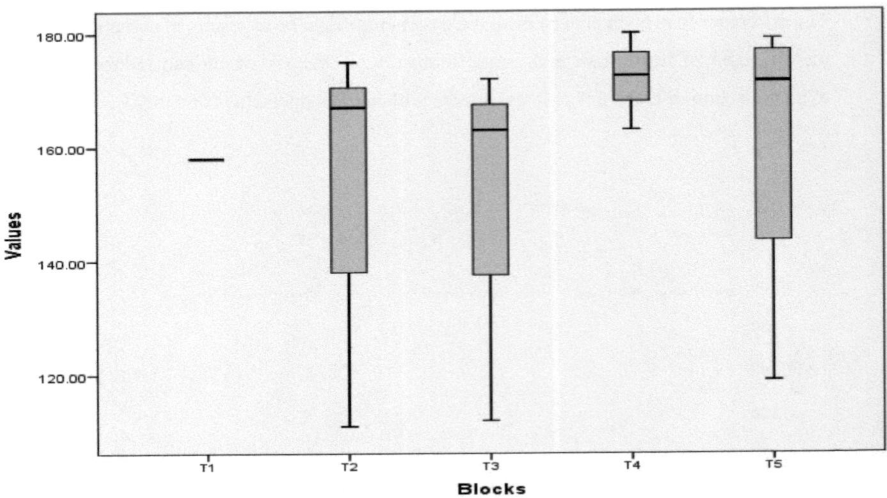

Figure 2: Box Plot showing values of different treatments

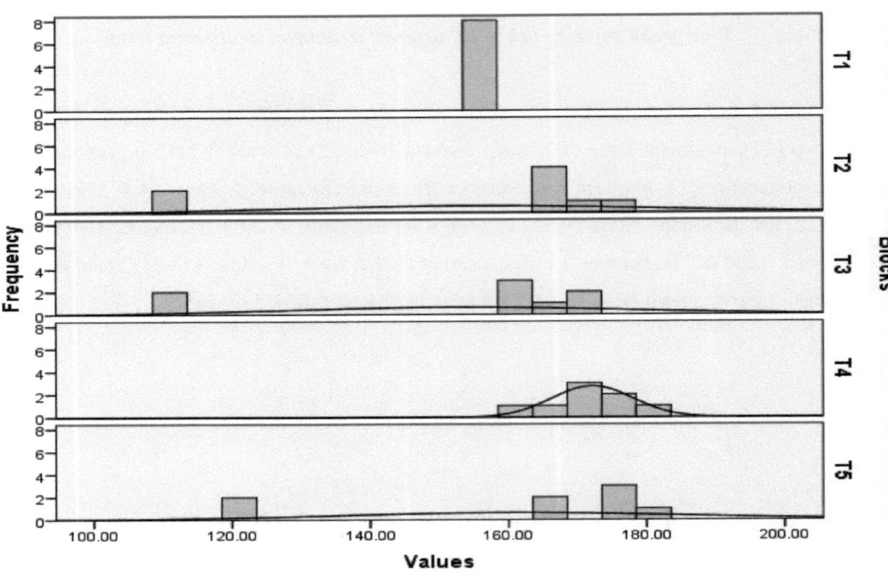

Figure 3: Histogram showing values of different treatments

Figure 2 depicts that treatment one (T1) has the same values from week 1 to week 8. Thus, have standard deviation of zero. Treatments T2, T3 and T5 are negatively skewed and have high standard deviation, which is also shown in Table 2. Treatment four (T4) is normally distributed and has small standard deviation. The box plot depicts five numbers, the extreme values (highest value and lowest value), the 1st and 3rd quartiles and the median. Figure 3 is the histogram, which shows the modal values and the normal curve and shows the skewness and peakedness. The bar chart of the mean and standard deviation of the blocks and weeks are displayed using multiple bar charts as depicted in figures 4 and 5 respectively.

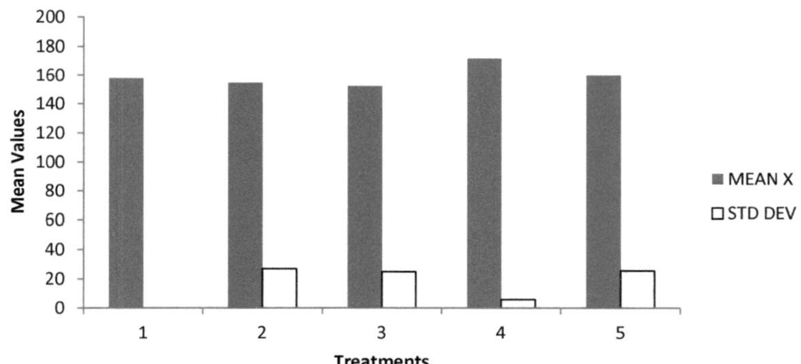

Figure 4: Bar Chart showing mean and standard deviation of different treatments

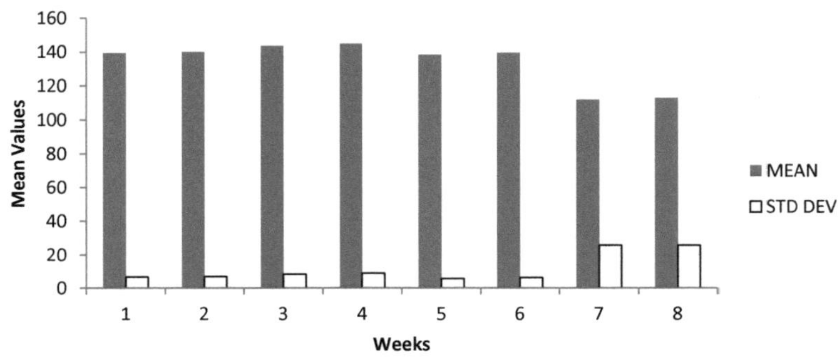

Figure 5: Bar Chart showing mean and standard deviation of different weeks

***Table 2: Randomized Complete Design (RCD): Two-way Analysis of Variance* (ANOVA)**

Source of Variation	Sum of Squares	Df	Mean Square	F	P-value
Treatments	1879.750	4	469.938	2.821	0.044
Weeks	9548.000	7	1364.000	8.188	0.000
Error	4664.250	28	166.580		
Total	1033702.000	40			
a. R Squared = 0.995 (Adjusted R Squared = 0.994)					

Table 2 shows the ANOVA summary table for the randomized complete block design and it shows that the null hypothesis is rejected, since the p-values 0.044 and 0.000 for treatments and weeks are less than the level of significance 0.05. This concludes that at least one of the treatments has a different mean and at least one of the weeks has a different mean. This implies that the mean values of the treatments are not the same at 5% level of significance. Since the test is significant a post hoc test is conducted which helps to classify homogeneous mean in the same subset. The post hoc test is carried out using the Duncan comparison test. Table 4 and Table 5 show the result of the post hoc test for treatments and weeks respectively.

Table 3: Post Hoc Test: Duncan Comparison Test for Blocks

	Subset	
Blocks	B	A
3	152.5000	
2	154.6250	
1	158.0000	
5	160.2500	160.2500
4		172.1250
P-value	0.283	0.076

Table 4: Post Hoc Test: Duncan Comparison Test for Weeks

	Subset	
Weeks	B	A
7	132.6000	
8	133.6000	
5		164.8000
6		166.2000
1		166.8000
2		167.6000
3		171.6000
4		172.8000
P-value	0.903	0.397

Table 3: It shows that treatments T4 and T5 means are not significantly different, so they belong to the same homogeneous subset "a", while T1, T2, T3 and T5 have homogeneous mean and are in the same subset "b". Note that subset "a" have higher means than subset "b".

Table 4 shows that week 1, 2, 3, 4 and 5 have homogeneous mean, so they belong to the same homogeneous subset "a", while week 7 and 8 have homogeneous mean and are in the same subset "b". Note that subset "a" have significant higher means than subset "b".

4.2 MORTALITY RECORDS

TABLE: 5 mortality records

Weeks	T1	T2	T3	T4	T5
1	-	2	1	-	-
2	1	4	2	2	3
3	2	3	-	-	2
4	1	9	3	3	-
5	3	3	6	1	4
6	1	4	-	2	-
7	-	2	2	-	1
8	4	3	1	4	1
No of Mortalities	12	30	15	12	11
Of crabs	30	30	30	30	30
No of crabs					
Survivals	18	0	15	18	19

4.3 DISCUSSION

The mortality rates throughout the course of the experiment were recorded. T**1** and T4 have the same mortality rates 40.00%, T2 have the highest mortality rates 100.00% while T3 have the second highest mortality rate 50%, and finally T5 have the highest survival rate with the lowest mortality rate of 36.67%.

Thus, this table depicted that the total survival rate is equal to 46.67%

Therefore, percentage of mortality rate = $\frac{\text{Total mortality x 100}}{\text{No of crabs}}$

In table 5, the high mortality rates recorded during this experiment in an open air was due to access to direct sunlight and water logged in the pen of *Cardisoma armatum* as result of heavy rainfall. Access to direct sunlight raised the temperature beyond the optimum temperature require by crab thereby led to high evaporation of water from crab body leading to loss of weight and also make the crab in-active in response to feed intake according to (aqualandpetsplus.com; 2007) stated that *Cardisoma armatum* do well at temperature75 – 85F. Exposure to heavy rainfall led to high mortality rates as the units were flooded, this is in agreement with the findings of (Matt Clarke; 2005,11-14) who posited that juvenile and adult crabs spend most of their time on dry land. According to (Adamezewske*et al;* 1997) many species actively forage on land and several species become semi-terrestrial. *Cardisoma armatum* being a land crab cannot survival in heavy flood more so; Land crab is a common name for multiple species of true crabs adapted for terrestrial existence according to (Encarta, 2009).

CHAPTER FIVE

CONCLUSION AND RECOMMENDATION

5.1 CONCLUSION

The project work was carried out and performed efficiently and the result was observed and discussed upon. According to the project work, it was discovered that *Cardisoma armatum* does not adapt to watery area for a long period of time, it was discovered that they tend to spend most of their life in marshy or wet environment. During the project work, it was also discovered too that they do not breed in captivity and also change their carapace when not only in captivity. Also, it was discovered that these organisms are carnivores, they tend to feed on each other especially on their mutilated parts, it was observed that *Cardisoma armatum* do not have a defined feeding habit, they feed on anything that comes their way, culturing *Cardisoma armatum* requires total attention due to their aggressiveness

5.2. RECOMMENDATION

Due to the importance of this project and the need in culturing crabs for commercial and subsistence uses, it can thus be recommended that;

1. Further research should be done on the improvement of culturing of crabs for better output.

2. *Cardisoma Armatum* should be cultured in fibre tanks due to their aggressive nature to avoid escape

3.*Cardisoma Armatum* should be cultured during the dry season, and if it is to be cultured during the raining season, it is should be provided with shield to avoid direct sunlight and rainfall to avoid water evaporation from the body of the crab and flood because they do not spend a long period of time in water.

REFERENCE

Abby-Kalio, N.J. (1982): Notes on crabs from the Niger Delta. The Nigerian field, 47(1- 3): 22-27 2. Hart, A.I. and Chindah, A.C. (1998): Preliminary study on the benthicmicro fauna associated with different micro habitats in mangrove forest of the bonny estuary, Niger Delta, Nigeria. Arch. Hydro Biol. 40:9-15 3.

Adamezewska ,A.m, van Aardt, W.j and moris , (1997).Role of lungs and gills in an African freshwater crabs, potamonauteswairew in gas exchange with water, with air and during exercise .Journal of crustaceans Biology 17,566.608.

Anon. 2006. Australian Prawn Farming Manual. Health management for profit. The State of Queensland, Department of Primary Industries and Fisheries. 157 pp.

Anon. 2007.Guidelines for constructing and maintaining aquaculture containment structures. The State of Queens land, DOPIAF. 40 pp. Quinitio, E.T. & Lwin, E.M.N. 2009. Soft-shell crab farming. SEAFDEC. 21 pp.

Anon. 2010. Land crab fishery management plan. Northern Territory of Australia. Arriola, F.J. 1940. A preliminary study of the life history of cardisoma armatum (Forskal). The Philipp. J. Sci.,73:437-454

Baliao, D.D., De Los Santos, M.A. & Franco, N.M. 1999. Pen culture of land crab in mangroves. SEAFDEC Aquaculture Extension Manual No. 26. 10 pp

Bell, J.D. & Gervis, M. 1999. New species for coastal aquaculture in the tropical Pacific – constraints,prospects and considerations. Aquacult. Int., 7(4): 207–223. Chakraborty, A., Otta, S.K., Joseph,
B., Kumar, S., Hossain, M.S., Karunasagar, I., Venugopal, M.N. & Karunasagar, I. 2002. Prevalence of white spot syndrome virus in wild crustaceans along the coast of India. Curr. Sci.,82(11): 1392–1397.

Bright,D., & C. Hogue. 1972. A synopsis of burrowing land crabs of the world and list of their arthropod symbionts and burrow associates. Contributions in science. No 220. Availableonline
(PDF)

Chen, L.-L., Lo, C.-F., Chiu, Y.-L., Chang, C.-F. & K ou, G.-H. 2000. Natural and experimental infection of white spot syndrome virus (WSSV) in benthic larvae of land crab cardisoma armatum. Dis. Aquat. Org., 40(2): 157–161.

Cowan, L. 1980. Crab farming in Japan, Taiwan and the Philippines. Information Series QI84009. Report Number 0727–6273. Queensland Department of Primary Industries.85pp. Daw12345678 ,"My rainbow crab eating watermelon,"ed. Youtube, 2010.

Cuesta et al 2002, Luppi TA, Rodriguez A Spivak ED(2002) Morphology of the megalopal stage of Rhithro Panopeus harrissi(Gould 18,4) (Crustacea Decap[da, Brachyura, Xanthidae) en la peninsula Iberica.Bol Inst Esp Oceanography 7:149-153

Dat, H.D. 1999. Description of land crab (Cardisoma spp.) culture methods in Vietnam. In C.P. Keenan & A. Blackshaw, eds. land crab aquaculture and biology. Proceedings of an intenational scientific forum held in Darwin,21–24 April 1997. pp. 67–71. ACIAR Proceedings 1999 of Conference

Encarta, 2009, the adult life of cardisoma Armatum.

Feliciano,C 1962. Notes on the biology and economic importance of the land crab of Puerto Rico. Spec. Contribution inst. Marine Biol. Univ.Puerto rico. 29pp

Gifford C (1963) Some observations on the general biology of the land crab in South Florida, Biological Bulletin 123: 207-233.

Hameed, A.S.S., Balasubramanian, G., Musthaq, S.S. & Yoganandhan, K. 2003. Experimental infection Of twenty species of Indian marine crabs with white spot syndrome virus (WSSV). Dis. Aquat. Org., 57(1–2): 157–161.

Haryanti, Sugama, K. & Nishijima, T. 2003. Diversity of bacteria isolated from crustacea larvae and their rearing water. J. Ocean Univ. Qingdao (Engl. Ed. Mar. Sci.), 2(1): 49–52.

He, X., Wang, C., Chen, K., Liu, Y. & Liu, J. 1992. Study disease on the sacculinary of Scylla. J. Zhanjiang Fish. Coll./Zhanjiang Shuichan Xueyuan Xuebao, 12(2): 41–45.

Herbst, 1794 Cardisoma carnifex– red-claw crab: found in Indo-Pacific coastal regions

Herklots and Rathbun.branchoyuran decapods crustacean callinectus latimanus,Association 3, 37-40 33 okafor, f.c,1988 the ecology of sudanonautes africanus h. Milne Edwards

Kanazawa, A. 1981. Penaeid nutrition. In G.D. Pruder, C. Langdon & D. Conklin, eds. Proceedings of the second international conference on aquaculture nutrition. Biochemical and Physiological Approaches to Shellfish Nutrition, pp. 87–105. Louisiana State University USA.

Kanchanaphum, P., Wongteerasupaya, C., Sitidilokratana, N., Boonsaeng, V., Panyim, S., Tassanakajon, A., Withyachumnarnkul, B. & Flegel, T.W. 1998. Experimental transmission of white spot syndrome virus (WSSV) from crabs to shrimp Penaeus monodon. Dis. Aquat. Org., 34(1): 1–7.

Kiatpathomchai, W., Jaroenram, W., Arunrut, N., Gangnonngiw, W., Boonyawiwat, V. & Sithigorngul, P. 2008. Experimental infections reveal that common Thai crustaceans are potential carriers for spread of exotic Taura syndrome virus. Dis. Aquat. Org., 79(3): 183190.

King, I, Childs. T.,Dorsett, C. Ostrander, J.G. and Monsen, E.R.(1990): Shellfish: Proximate composition, minerals, fatty acid and sterols. Jour. Amer. Dietetic Ass. 90:677-685 8.
Latreille, Cardisoma Guanhumi 1825-Blue land crab or giant land crab: found in Western Atlantic coastal regions. Capture and commercialization of blue land crabs.Biol Ecol, 65-47-65

Lo, C.F., Ho, C.H., Peng, S.E., Chen, C.H., Hsu, H.C., Chiu, Y.L., Chang, C.F., Liu, K.F., Su, M.S., Wang, C.H. & Kou, G.H. 1996. White spot syndrome baculovirus (WSBV) detected in cultured and captured shrimp, crabs and other arthropods. Dis. Aquat. Org., 27(3): 215–225.
Matt Clarke (2005 -11-14)." Rainbow crab, Cardisoma armatum". Practical fish keeping.D. Waren."cardisoma Armatum". Retrieved 2007-07-16
N.M. 1999. Pen culture of mud crab in mangroves. SEAFDEC Aquaculture Extension Manual No. 26. 10 pp.

Otta, S.K., Shubha, G., Joseph, B., Chakraborty, A., Karunasabar, I. & Karunasagar, I. 1999. Polymerase chain reaction (PCR) detection of white spot syndrome virus (WSSV) in cultured and wild crustaceans in India. Dis. Aquat. Organ, 38: 67–70.

Oyekanmi, A. (1984): Outline of food analysis. 1st Ed. MacMillan. London. Pp14-72 4. Hall, D. Lee, S.Y. and Meziane, T. (2006): Fatty acid as tropic tracer in experimental estuarine food chains. Tracer transfer Jour.Exptal.Poniedzialek. (2011). Crustaceans-Rainbow crab Available: http//doroorod.blogspot.com/2011/05/crustaceans-Rainbow-crab. Html

Poole, S., Mayze, J., Exley, P. & Paulo, C. 2008. Maximizing revenue within the NT crab fishery by enhancing post-harvest survival of mud crabs. Report Number Project No.2003/240, FisheriesResearch and Development Corporation, A. 1–154

Ruscoe, I.M., Shelley, C.C. & Williams, G.R. 2004. The combined effects of temperature and salinity on growth and survival of juvenile land crabs (cardisoma armatum Forskal). Aquaculture, 238 (1–4): 239–247.

Scot C. Nelson, et al: 2006.Musa species(banana and plantains) species profiles for percific island and Agroforestry

Smith, 1870 Cardisoma crassum– mouthless crab: found in east Pacific coastal regions

Skonberg, D.I. and Perkins, B.L. (2002): Nutrient composition of green crab (Carcinus maenus) legmeat and clawmeat. Food Chem., 72:401-404 9. USDA (2003): US Department of Agriculture,National nutrient data base for standard reference, Release 16: Nutrient data laboratory Home page. Available hppt://www.nal.usda.gov/fnic/foodcomp/search 10.

Smallegange, I.M. and Van Der Meer, J. (2003): Why do shore crabs not prefer the most profitable mussels? Jour. Anim. Ecol. (72): 599-607 7.

Taissoun,E.N. 1974. El cangrejo de tierra (Cardiosoma Armatum) Venuzuela.Boletin del centro deinvestigaciones Biologicas, Vol . 10.

W.W Burggren and B.r Mc Mahon,(cds.),1988. Biology of the land crabs xii,1-479, text-figs(Cambridge university press New York)

Yano, Y., Kaneniwa, M., Satomi, M., Oikawa, H. & Chen, S.S. 2006. Occurrence and density of Vibrio parahaemolyticus in live edible Crustaceans from markets in China. J. Food Protect, 69(11): 2742–2746.

Zhen, J. & Shao, Y. 2001. Experimental study on thermal environment in over-wintering greenhouse for cardisoma armatum. J. Zhejiang Ocean Univ. (Nat. Sci.)/Zhejiang Haiyang Xueyuan Xuebao, 20(3): 205–208.